Geometric Patterns

from
Tiles & Brickwork

Robert Field

Tarquin Publications

Fountains Abbey, Yorkshire

All the photographs and illustrations are by the author with the exception of the three photographs on the bottom of page 36 which were taken by Anthony Beeson to whom many thanks are due for his permission to use them.

© 1996:   Robert Field   Tarquin Publications
I.S.B.N.: 1 899618 12 0   Stradbroke
Design: Magdalen Bear   Diss
Printing: Ancient House Press, Ipswich   Norfolk. IP21 5JP

England

  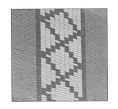 

"Seek no dull repetition
But delight in what can be!"

Bricks and tiles made from fired clay are two of the most universal and ancient building materials known to man. The technique of firing clay to make it stronger and more durable was first discovered around 4000 to 4500 years ago and was used by both the Mesopotamian and Indus Valley civilisations. Two important qualities of both bricks and tiles are that the unit sizes are convenient to handle and that as many as required for the task in hand can be manufactured. The repetitive nature of the units is more an artistic opportunity than a restriction, as we shall see throughout this book.

The clay is dug from the earth and mixed with water. It is then pressed into a mould and allowed to dry before being fired. A standard mould can be used to produce many thousands or even millions of broadly identical items. However, there is always a certain unpredictability about the result, even using clay from different parts of the same pit. The colour of each brick or tile depends both on the mineral content of the clay and also on the temperature and the duration of the firing. These variations have been exploited by builders, designers and architects over the centuries to provide visually exciting exteriors to their buildings.

Although this book is mostly concerned with the geometric patterns made by bricks and tiles, of course practical considerations matter enormously. No wall or building is any use, no matter how beautiful, if it falls down, develops cracks or lets the rain in. The attractiveness of the patterns is enhanced, not reduced by the need to construct buildings which will last and fulfil their purpose.

Such a book as this cannot be an exhaustive study, but hopefully will serve as a starting point for the reader to explore his or her own environment in search of the wonderful and ingenious geometric patterns that buildings can provide. The patterns discovered or suggested by tiles and brickwork can readily serve as a basis for designs for other artistic endeavours. Needlework, patchwork, marquetry, collage and fabric design are just some of the craft activities which spring to mind.

I hope you derive as much pleasure from them as I have gained in compiling this book.

Robert Field.

Brick Bonds

In order to make a strong and durable wall, the bricks have to be laid so that the joins in one course do not come vertically above or below the joins in neighbouring courses. Where the long side of the brick shows it is called a *stretcher*. Where the short side shows it is called a *header*. Many different patterns of headers and stretchers are possible and they are known as *bonds*. Two of the commonest bonds are shown below.

Header

Stretcher

Flemish bond, Bristol

English bond, Farnham, Surrey

These photographs show just some of the remarkable range of colours and textures that brick walls are able to offer to the architect. The use of different clays and different bonds as well as the random effects of different firing conditions all combine to add richness and variety to the surfaces. Even the *mortar* which bonds the bricks together can be made in a range of colours to complement or contrast with the chosen bricks.

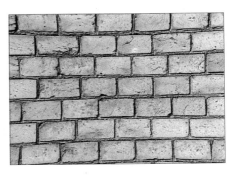

Cavity Walls

The patterns seen on the outside of walls are very much related to what happens inside. Modern requirements for damp proofing and insulation demand that brick walls for buildings are built with a cavity. The inner leaf is generally constructed with insulation blocks and so the outer leaf is only the thickness of a brick placed lengthways. Only stretchers show and this method of bonding is known as *stretcher bond*. Virtually all modern buildings use it. The two leaves are linked with metal *ties* at suitable intervals.

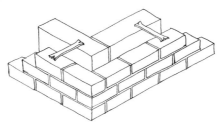

Solid Walls

Bricks are twice as long as they are wide and in thicker or 'solid' walls this means that they can be bonded in a variety of different criss-cross ways. There are many different ways of doing this and sixteen of the best known are listed on pages 7-9. A good bond is one with no vertical joints one above another and where few bricks have to be cut.

Flemish Bond

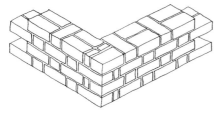

English Bond

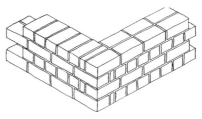

The two solid walls illustrated above are called 'one brick' or 'nine inch' walls. They show how the bricks are laid in two of the commonest bonds to produce such a wall. The size of bricks has varied over the centuries, but they were traditionally made to be just less than nine inches long, four and a half inches wide and two and a half inches high. Although nowadays brick sizes are metricated and a standard brick is 215 x 102.5 x 65mm, the former descriptions of 'four and a half inch' and 'nine inch' are still commonly used for wall thicknesses.

Walls can also be 'one and a half' or 'two' bricks thick or even thicker, but with most modern buildings making use of a steel or reinforced concrete framing it is seldom that walls thicker than 'one brick' are required.

Since individual walls and buildings are seldom exact multiples of a brick length, the adjustment is made with smaller cut pieces worked in near to the corners. They are known as *closers*, and the arrangement of the closers can themselves create an attractive pattern.

The Flemish Bond and its variations

Flemish Bond
Each course has alternate headers and stretchers. The headers are centred on the stretchers above and below.

Flemish Garden Wall Bond
This variation has not one stretcher between the headers, but three. It is weaker because there is less cross-bonding, but was popular in Sussex and is sometimes known as Sussex Garden Wall Bond.

Dutch Bond
In this variation the headers are moved by half a brick and are no longer centred under the stretchers.

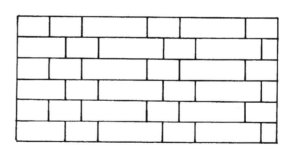

Monk Bond
This variation has two stretchers between the headers. It looks very distinctive when the headers are a different colour from the stretchers.

The English Bond and its variations

English Bond

This is a very strong bond which can be seen in use very commonly. Alternate courses have headers only and stretchers only.

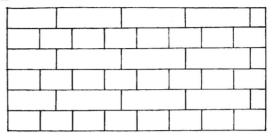

English Cross Bond

Alternate courses of stretchers are moved over by half a brick. The effect of this variation is to give the wall a strong stepped pattern of vertical joints.

Double English Cross Bond

As the name suggests, two courses of headers alternate with two courses of stretchers. It has the stepped effect of the English cross bond, but in one direction, not two.

Dearne's Bond

In this variation the stretcher courses are laid edge-on. This creates a cavity and fewer bricks are required. It was used in walls for gardens, boundaries and outhouses.

The Stretcher Bond and its variations

Stretcher Bond

Each course consists only of stretchers which overlap by half a brick. It is used for cavity walls almost exclusively nowadays. It is rarely used for solid walls because there is no cross-bonding.

Raking Stretcher Bond

In this variation, the stretchers overlap by a quarter of a brick rather than the usual half a brick. It can be used for cavity walls but not for solid walls as there is no cross-bonding.

Flemish Stretcher Bond

This variation overcomes this lack of cross-bonding by introducing single courses of Flemish bond at regular intervals into the stretcher bond.

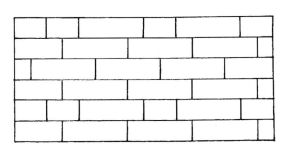

English Garden Wall

In a similar way, this variation introduces a course of headers, as in English bond, every three or five courses of stretchers.

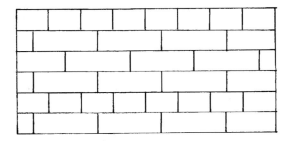

Miscellaneous Bonds

Header Bond

Only the headers show and it is a strong bond which can be used for slightly curving walls. Look for it on bay windows.

Stack Bond

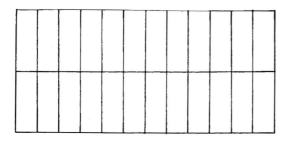

The courses of bricks are stacked end to end. Since it leaves continuous vertical joints, it cannot be used for a load bearing wall. However, it is used to good decorative effect in buildings with a steel frame.

Rat Trap Bond

All the bricks are laid edge-on in a Flemish bond pattern. This creates cavities between the stretchers. It uses fewer bricks and is therefore cheaper.

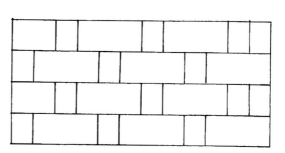

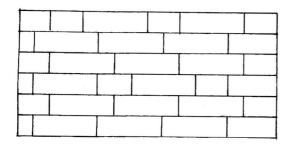

Mixed Garden Bond

This bond is a variation of the Flemish stretcher bond in which the headers are arranged so that they avoid the formation of vertical columns.

Liberal Club, Farnham, Surrey

Special Orders

In general, bricks are mass-produced objects and they are only available in standard thickness, generally 65mm. However, it is possible for an architect with a client who is willing to pay the extra cost, to design a building with a special brick which is ordered and made exclusively for it.

One famous architect Sir Edwin Lutyens (1869-1944) was fond of thinner bricks and for the facade of the Liberal Club in Farnham, he ordered some special reds. They were to be laid in Flemish bond and were two-thirds the thickness of a normal brick. However, probably because of the expense, the side walls were built of normal standard bricks of a different colour.

The photograph shows how the join was achieved so that the two types would be firmly bonded together.

Liberal Club, Farnham, Surrey

Roman brickwork at Pompeii

Looking back to the Romans

Before the days of coal, oil, gas or electricity, the firing of bricks could only be achieved by burning wood in kilns. This was an expensive process and difficult to organise as sufficient fuel had to be gathered together to complete the firing at one attempt. Bricks cannot be partly fired one day and then finished later when more fuel becomes available.

Most types of Roman bricks were square and their sizes were roughly based upon the unit of the foot. The names were as follows:

Pedalis	One foot square (300 x 300 mm),
Sesquipedalis	One-and-a half foot square (450 x 450 mm),
Bipedalis	Two feet square (600 x 600 mm),
Bessalis	Two-thirds of a foot square (200 x 200 mm), and
Lydion	Rectangular, one by one-and-a-half feet (300 x 450 mm).

Roman bricks were thinner than modern bricks, averaging only about two inches or 50 mm in thickness.

The Romans rarely built walls completely of brick but cut the square bricks diagonally in half and used them as facings for walls which were then infilled with rubble and mortar. Another use for bricks was as a bonding course every so often to help stabilise walls otherwise built of stones. Bricks were commonly used to construct the pillars to support the floor above a hypocaust.

Merida, Spain

The new Roman Museum in Merida in Spain has been built of bricks which were specially manufactured in a Roman style.

The imaginative use of these special thin bricks has given both the inside and the outside an appearance which is very appropriate for such a building.

Merida, Spain

Brick Nogging

When a building has a timber frame, it is the frame itself which carries most of the weight of the walls and the roof. It also provides most of the lateral strength. The area between the wooden beams is frequently filled in with standard bricks. In these circumstances, the bricks can be laid in a more decorative way than could ever be permitted in a load bearing wall. This infill is known as brick nogging.

It is common to see such bricks arranged in a herring-bone pattern or in one of the other designs on the opposite page.

Singleton, Sussex

The Sun Hotel, Canterbury

Horizontal Stack Bond

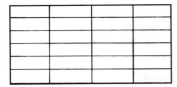

Vertical Stack Bond

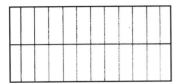

Vertical Basket Weave

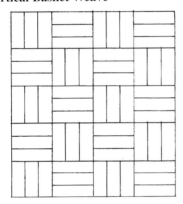

Vertical Herring-bone

Diagonal Herring-bone

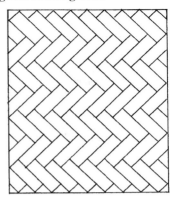

Double Herring-bone

Diagonal Basket Weave

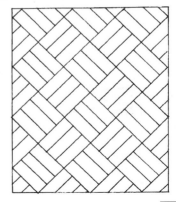

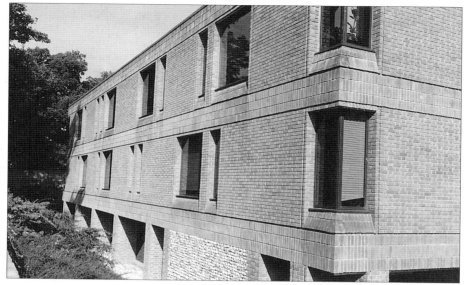
Chichester, Sussex

Steel Framing

Nowadays many buildings, both commercial and residential, have a steel structure which carries the load and the outside brick walls are simply a decorative and weather-proof skin. In these circumstances, areas of the brickwork can be laid in the weaker bonds such as stack, basket weave and herring-bone shown overleaf. Sometimes these patterns are picked out in colour but often the change of texture is sufficient.

Montpelier House, Brompton Road, London

Weybridge, Surrey

Docklands, London

Knightsbridge, London

Selsdon, Surrey

The Vyne, near Basingstoke, Hampshire

By choosing clay from different locations or consciously varying the firing time, a manufacturer can produce bricks with very different colours and textures. By inter-mixing distinctive types in a systematic way, regular patterns can be easily produced. This technique was very popular in Tudor times and it was used to produce what are known as *diaper patterns*. These are usually in the form of a network of lozenges.

A closer view of the house above

Chichester, Sussex

Building in Flemish bond using red bricks as stretchers and vitrified blue-black bricks as headers, produces an attractive chequered appearance.

Alresford, Hampshire

Patterns in Stretcher Bond

Variety can be added to the ubiquitous and economical stretcher bond by the use of different colours and textures. All of these patterns have been achieved by using bricks of two different colours.

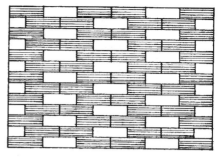

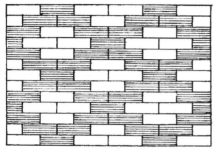

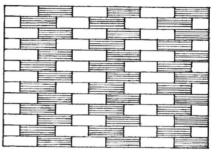

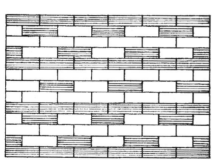

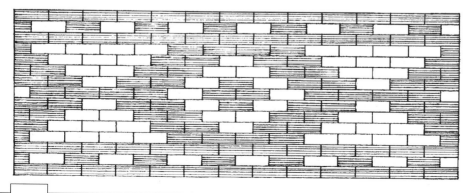

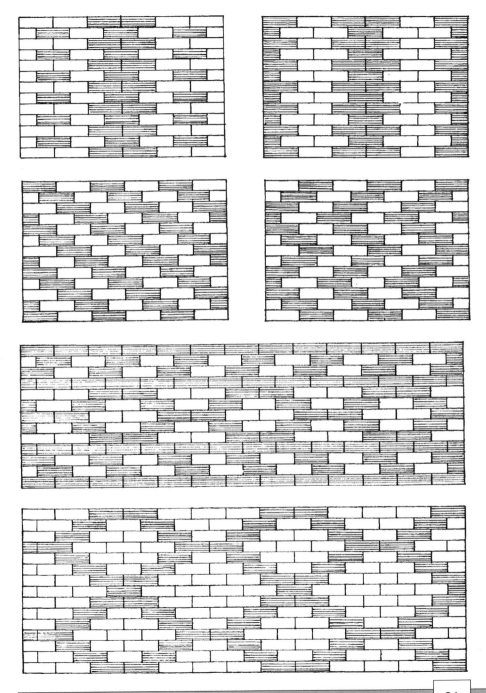

Patterns in Header Bond

Header bond shows only headers and all of these patterns have been achieved using bricks in just three different colours.

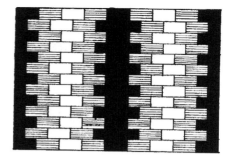

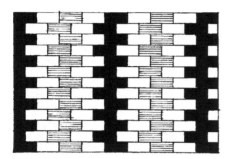

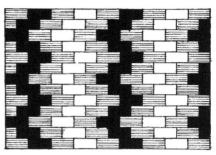

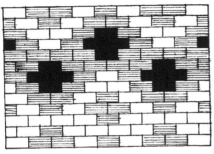
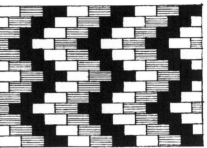

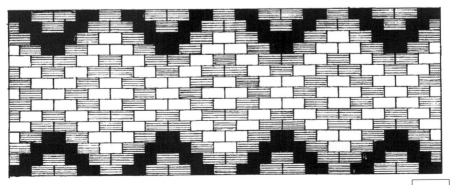

23

Patterns in English Cross Bond

The characteristic of this bond of moving alternate courses of stretchers over by half a brick permits the design of many interesting patterns. All of these patterns have been achieved using bricks in three different colours.

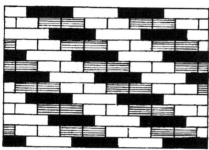

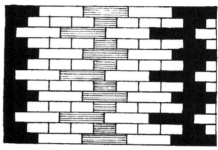

Patterns in Flemish Bond

Flemish bond shows equal numbers of headers and stretchers and can be developed into a range of patterns in either two or three colours. All of these patterns have been achieved using bricks in three different colours.

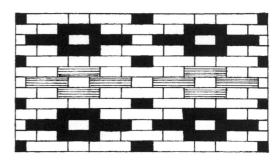

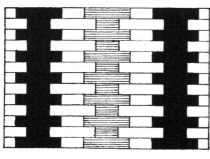

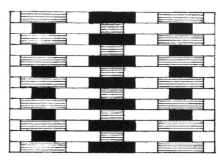

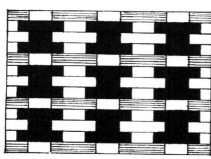

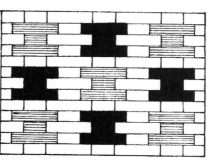

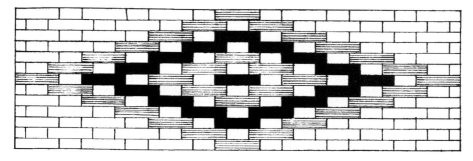

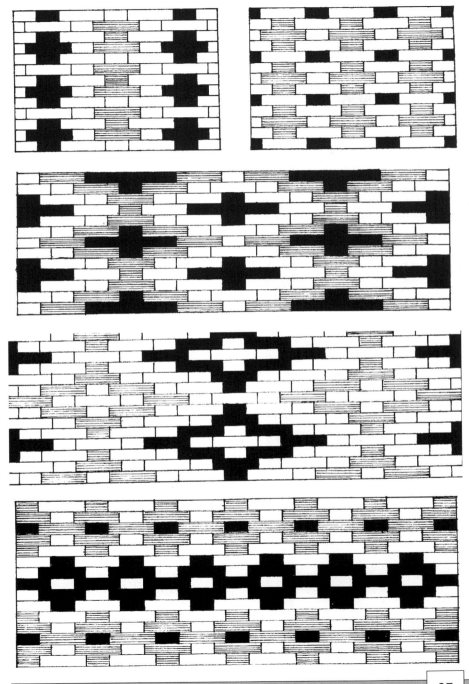

Ironbridge, Shropshire

In Victorian times the use of coloured bricks became extremely popular. During this period many of the patterns listed on pages 20 to 27 were devised and used. They were very much an integral part of the design and conception of the buildings.

Reading University, Berkshire

All Saints, Margaret Street, London

Ironbridge, Shropshire

Wapping, London

The use of patterned brickwork to introduce variety into the urban landscape has again become very popular; in particular, in the redevelopment of London Docklands.

Leeds, Yorkshire

Docklands, London

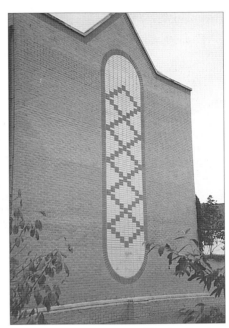

Wapping, London.

Whitechapel, London

Docklands, London

Breaking the Surface

Although most bricks are laid to present a flat, rain-resistant surface, many interesting designs and patterns can be created if some bricks are allowed to protrude.

All the photographs on this page were taken in Tozeur, an oasis city in Tunisia. In a dry, hot climate, a local and very distinctive style of brick decoration has evolved where scarcely any bricks are allowed to form flat surfaces at all!

Petersfield, Hampshire

In Britain the primary requirement of keeping out the rain cannot be ignored and such excesses could not be contemplated. However, protruding bricks are used in both older and newer buildings. Since weathering can be a problem, these styles of decoration are most frequently confined to protected areas beneath a gable.

Salisbury, Wiltshire

Co-operative Bank, Prescot Street, London

Farnham, Surrey

For some designs, ordinary bricks are laid at an angle of 45° to the surface. For others, special carved bricks are made and then raised patterns are created by building them into the walls.

Farnham, Surrey

Reading, Berkshire

An elaborate example of carved and protruding brickwork, well protected from the weather by an overhanging roof.

Weymouth, Dorset

Farnham, Surrey

Inns of Court, London

Tudor chimneys and their Victorian copies provide perhaps the most dramatic examples of patterned brickwork. Their builders did not worry too much about weathering!

Guildford, Surrey

The need for well ventilated spaces for car parks and for screens to hide such things as dustbins and central heating boilers has allowed architects to produce many buildings with striking geometrical designs.

While brick can be used for some of these designs where there is a steel or concrete framework to carry the load, the patterns are usually generated by tessellating specially formed concrete blocks.

Bristol

Bristol

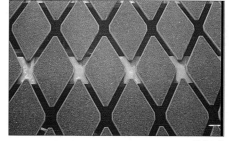

Bristol

London

Canterbury, Kent

Bond Street, London

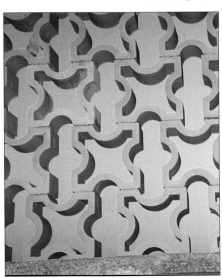

Boulogne, France

Using Bricks for Paving

Over the last few years the use of brick-shaped concrete blocks or *paviers* has become important for outdoor areas such as patios, paths, walkways, roads, pedestrian precincts, car parks and large open areas. Paviers are extremely strong and durable and can safely bear the weight, even of heavily-laden lorries. They are manufactured in a variety of colours and can be laid in many attractive and ingenious patterns.

Wapping, London

Wolverhampton

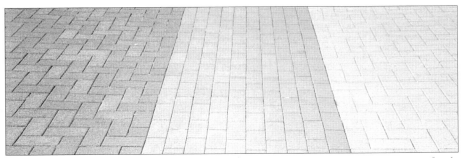

Leeds

Leeds

Bricks made from fired clay are weaker than paviers but they have been used since Roman times for paving rooms within houses and courtyards. The illustrations below show examples of surviving Roman floors where thin tile-like bricks have been laid in the traditional herring-bone pattern. Such work is known as Opus Spicatum.

Villa Jovis, Capri, Italy

Wroxeter, Shropshire

Il Campo, Siena, Italy

Here is a situation where fired brick is used to pave a large area in the open. It is the well-known piazza in Siena where the Palio, the yearly horse race, takes place.

Il Campo, Siena, Italy

Both concrete paviers and fired floor bricks can be laid in the same bonds as bricks in vertical walls, but generally other patterns are used. These are the most common.

Stretcher

Herring-bone

Basket Weave

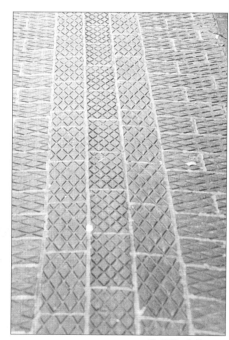

Paviers are sometimes impressed or incised with other patterns. On the right, the pattern is a deep cut network of diamonds.

Guildford, Surrey

Stack

Flemish

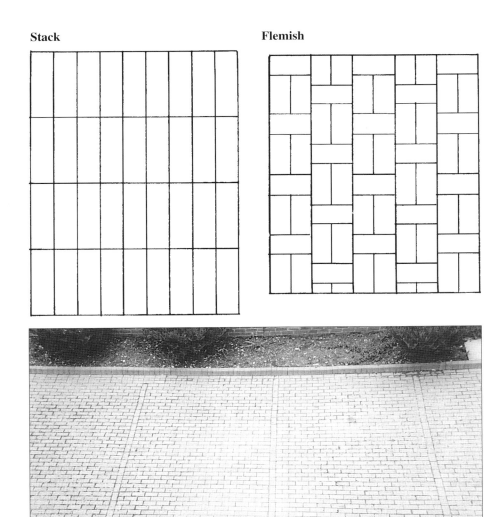

This car-park in Wapping, London has its parking bays marked out by single lines of bricks. The infilling bricks are laid in stretcher bond, cut to fit to those lines. The entry and exit roadway is indicated by the use of a basket weave pattern.

Concrete paving slabs in a variety of colours and shapes are widely used for domestic patios and paths as well as for civic and commercial areas. Although they can be made by hand and on a small scale, commercially manufactured slabs have the advantage of consistency of colour and texture. The use of a mechanical vibrator in the course of their production gives them extra strength and longevity.

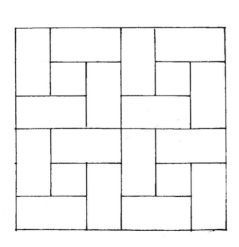

Dover

Oxford

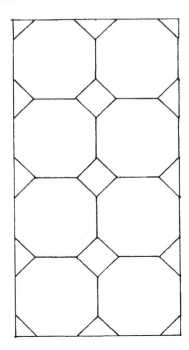

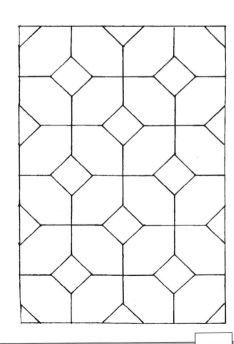

Roof Tiles

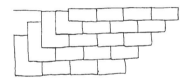

This view of Farnham in Surrey shows a variety of tiled roofs in different materials and styles. Some of the tiles are made of fired clay and some are slates. Virtually all are laid in what appears to be stretcher bond. However tiles are not bonded together with mortar, but overlap and are held in place by nails and their combined weight. They overlap for more than half their length and parts of the roof are three tiles thick.

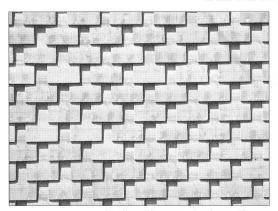

Generally, neighbouring tiles in the same row touch, but this dramatic pattern of wall covering has been achieved by leaving wide gaps between them.

Equihen-Plage near Boulogne, France

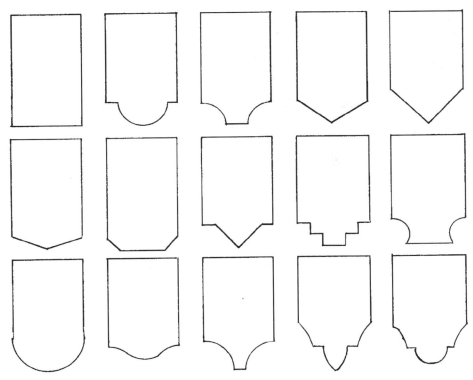

Farnham, Surrey

In order to create roofs with more interesting patterns, tiles with shaped lower ends can be introduced. Most of the designs above are used on tile-hung walls but sometimes rows or groups of a shaped tile of just one design are worked into a roof which is otherwise constructed of standard rectangular tiles. In other situations, two or more designs are combined to make an elaborate and intricate pattern.

In certain localities, particular local styles and combinations become favoured and fashionable and they then impart a distinctive character to the town or village.

Stradbroke, Suffolk

Pantiles are widely used in the Eastern Counties and give a strong rectangular pattern to the roofs of the area. The individual tiles are slightly curved and have a protrusion which hooks over the horizontal battens. No nails or mortar are required except at the ridge.

Roofs such as these also create very strong patterns, especially when seen in the normal conditions of bright sunlight and strong shadows. They are made from two different kinds of fired clay tile, a lipped flat plate called a tegula and a semi circular tapered half-pipe called an imbrex and have been in use since Roman times. It is a relatively cheap form of roofing because there is so little overlap.

Volterra, Italy

Agios Lavrendis, Greece

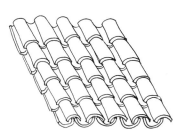

This kind of tiled roof is commonly seen in Greece, Italy and other Mediterranean countries. It consists of semi-circular tapered half-pipe lengths placed alternately upwards and downwards, thus forming a multitude of parallel channels for the rainwater and visually a powerful design of parallel lines. Since the overlap within both the channels and the ridges is small, roofs which use such tiles must be laid at a steep angle.

Assisi, Italy

Godalming, Surrey

Asbestos Tiles

Tiles made from a mixture of asbestos fibres and cement became quite widely used for farm and industrial buildings before the potential health hazard of asbestos was fully appreciated. The roof of the building above shows a typical diamond patterning. Interestingly, it closely echoes the pattern of the original Roman stone roofing slabs from High Ham Villa in Somerset, now on show at Taunton Museum.

High Ham Villa, Somerset

Slates

Slate is a widely used natural roofing material. It occurs when certain muds and clays have been baked under high pressure deep within the earth. It splits easily into flat sheets which can be cut and shaped into convenient and standard sized tiles, normally known as *slates*. Generally slates are a dark grey colour with blue, purple or green tints and are laid in a wide variety of different geometric patterns.

Slate Museum , Llanberis, Gwynedd

Slate Museum , Llanberis, Gwynedd

Weybridge, Surrey

51

Farnham, Surrey

Tile Hung Walls

In Britain, strong winds and driving rain are sufficiently common for most roofs of houses to be constructed from flat tiles or slates. Their ends may be shaped, but essentially they lie flat. Such tiles can also be hung on vertical walls to provide a waterproof and decorative surface.

In the upper photograph, hanging tiles have been used on the gables of the dormer windows so as to complement both kinds of tiles in the roof.

The lower picture shows how tiles can cover a complete wall. Tile-hung walls like this one can be built of a cheaper brick and therefore may not be as expensive to build as they seem.

Crondall, Hampshire

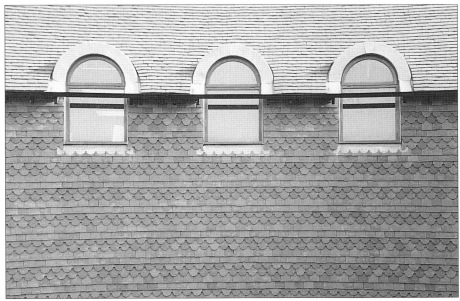

Tiles with shaped ends can be used sparingly as a dramatic focus in an otherwise plain roof or wall or widely in an exuberant display of pattern.

Crondall, Hampshire

Sometimes hanging tiles are used to add character to just the upper storey of a two storey house, but it is much more common to see them used as a decorative area above or between windows, especially on the front or street side of the house. When the walls on which they are hung are vertical, their function is entirely visual but they also can act as a steeply sloping roof to a bay window.

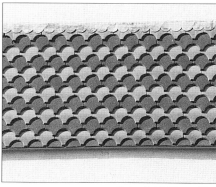

Petersfield, Hampshire

Bucks Horn Oak, Hampshire

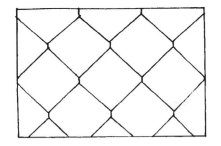

Merrow, Surrey

Farnham, Surrey

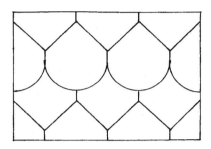

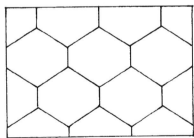

Canterbury, Kent

Salisbury, Wiltshire

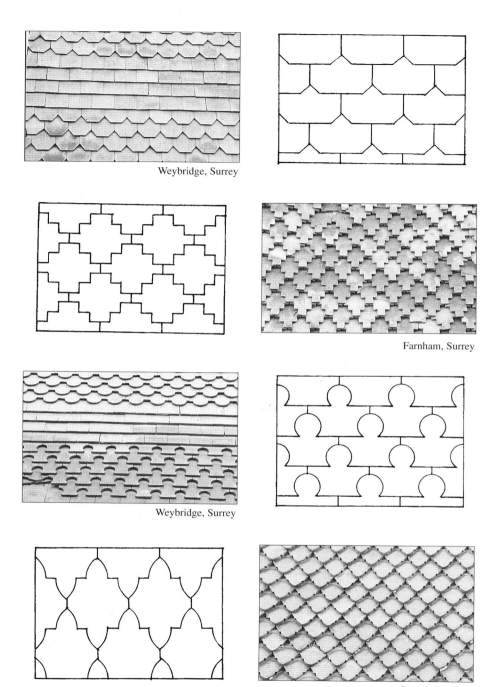

Weybridge, Surrey

Farnham, Surrey

Weybridge, Surrey

Canterbury, Kent

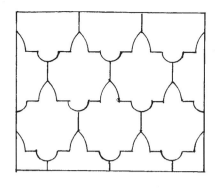

Elstead, Surrey

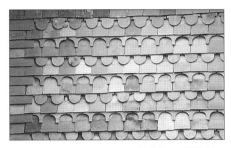

Salisbury, Wiltshire

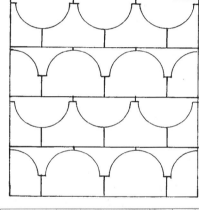

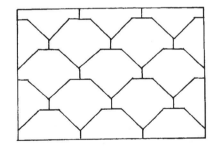

Petersfield, Hampshire

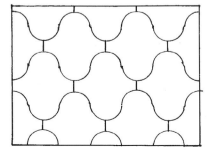

Petersfield, Hampshire

57

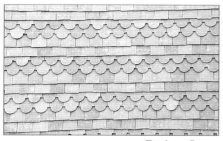
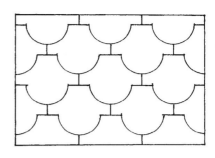

Farnham, Surrey

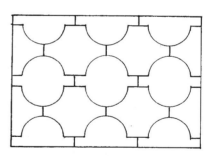

Weybridge, Surrey

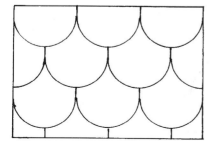

Crondall, Hampshire

Guildford, Surrey

Decorated Floor Tiles

When used for a floor, tiles do not overlap and are set firmly into a cement base. The ones on this page were made from a red clay and the designs were impressed on them while they were still wet with a wooden pattern. Those impressions were filled with another lighter coloured clay, thus creating a two-colour design. The tiles were then fired and glazed. Some of these tiles have symmetrical patterns on them; others are less symmetrical and are designed to be laid in larger sequences to complete the floor.

Winchester Cathedral

St. Andrew, Trent, Dorset

In Victorian times, when many new churches were built and older ones renovated, architects needed large numbers of new tiles for the floors. These were made by machine, but many of the designs and patterns looked back to earlier times, sometimes even to the medieval period. Some were exact copies of the hand-made tiles that they replaced, others were specially designed by the artists of the period. During this time of enthusiasm and invention some similarly magnificent floors were also laid in public buildings, such as town halls and post offices.

All Saints, Crondall, Hampshire

St. Peter and St. Paul, Godalming, Surrey

Simpler tiles in plain colours and standard shapes were also manufactured. They were either laid in designs which framed the more ornate and expensive decorated tiles or used to create patterned floors in their own right.

Jackfield Tile Museum, Ironbridge, Shropshire

St. Andrew, Trent, Dorset

St. Mary the Virgin, Silchester, Hampshire

Holy Trinity, Leeds, Yorkshire

Ridge Tiles

Roofs need to be sealed against the weather at the ridge and some architects have seized the chance to experiment with developing linear patterns there. Look upwards to see a rich and inventive use of repetitive elements.

Seldom are more than two different types of ridge tile used on the same building.

Reading, Berkshire

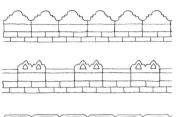

Reading, Berkshire

Weybridge, Surrey

Farnham, Surrey

Two designs from Petersfield, Hampshire

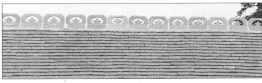

Farnham, Surrey

Weybridge, Surrey

Guildford, Surrey

Salisbury, Wiltshire

Weybridge, Surrey

Farnham, Surrey

Let us finish with two splendid examples of tiles and brickwork in normal suburban streets. Geometrical patterns are everywhere and if this book has stimulated you to look for them and to notice their richness, then it has succeeded in its objective.